Marvels of Space
50 Amazing Stories of the Universe

Sam Blackwood

Sameera Sandaruwan
No. B/8, Namadagala, Medamahanuwara
Kandy, Sri Lanka. 20940
sameerahotmail@gmail.com

DEDICATION

To my dearest little ones, Sandas and Sithev,
This book is dedicated to you, my stars up above.
May you explore the cosmos with wonder and glee,
And discover the mysteries of space that we can't yet see.

May you reach for the stars and chase your dreams so bright,
And find inspiration in the beauty of the cosmic night.
May you always remember to look up at the sky,
And know that the universe is vast, and you are not just a tiny guy.

So here's to you, my little astronauts in training,
May this book fuel your curiosity and keep you gaining
Knowledge of the marvels of space, both near and far,
And may you shine as bright as the brightest star.

CONTENTS

ACKNOWLEDGMENTS

Writing a book is never a solitary endeavor, and I could not have completed this work without the help and support of many individuals.

First and foremost, I would like to express my gratitude to my family, especially my curious little kids, Sandas and Sithev, who inspire me daily to learn and explore the mysteries of the universe. Your curiosity and wonder about space have motivated me to write this book, and I hope it will inspire you and other young minds to continue to explore the cosmos.

I would like to thank the team at OpenAI for providing me with the vast knowledge necessary to write this book.

I would also like to thank my editor, Sameera Sandaruwan, who provided valuable feedback and guidance throughout the writing process. Your attention to detail and constructive criticism have helped me refine my work and make it better.

Lastly, I would like to acknowledge the countless scientists, researchers, and astronomers whose hard work and dedication have helped us unravel the mysteries of the universe. This book is a tribute to your contributions to human knowledge and understanding of the cosmos.

Sam Blackwood

The First Moon Landing
Neil Armstrong's Historic Moment

In 1961, President John F. Kennedy challenged the United States to land a man on the moon before the decade's end. This challenge set in motion one of human history's most ambitious and complex missions.

Fast forward to July 20, 1969, and the world watched as Neil Armstrong, Buzz Aldrin, and Michael Collins embarked on the Apollo 11 mission to fulfil President Kennedy's challenge. Their spacecraft, the Saturn V rocket, blasted off Earth and travelled over 200,000 miles to reach the moon.

When the lunar module, named Eagle, separated from the command module, Columbia, Armstrong, and Aldrin began their descent towards the moon's surface. As they got closer, they encountered unexpected boulders and had to adjust their landing site quickly.

As the Eagle descended, alarms began to sound, indicating that the lunar module was running low on fuel. With only seconds of power remaining, Armstrong had to manually pilot the Eagle to a safe landing spot, avoiding even more boulders.

Finally, at 10:56 pm Eastern Standard Time, Armstrong stepped out of the lunar module and said, "That's one small step for man, one giant leap for mankind." He became the first human to set foot on the moon, with Aldrin following shortly after.

Armstrong and Aldrin conducted experiments and collected rock samples while Collins orbited the moon, waiting for his fellow astronauts' return.

After spending over 21 hours on the lunar surface, the Eagle launched back into space to dock with the Columbia. The astronauts then journeyed back to Earth, where they were greeted as heroes.

The first moon landing was a monumental achievement for the United States and humanity. It represented the culmination of years of hard work and determination by thousands of people, from the astronauts to the engineers to the support staff.

The mission also helped to advance our understanding of the universe and our place in it. The rocks and soil samples collected by the astronauts have been studied for decades, providing scientists with valuable information about the moon's composition and history.

Since the Apollo 11 mission, several other countries and organizations have sent missions to the moon, further expanding our knowledge of Earth's only natural satellite. However, the first moon landing remains a historic moment that will forever be remembered as a symbol of human ingenuity and perseverance.

So there you have it, the story of the first moon landing and Neil Armstrong's historic moment. It's an incredible achievement that inspires people of all ages to reach for the stars and explore the unknown.

Revealing the Secrets of the Universe
The Hubble Space Telescope

The Hubble Space Telescope is a remarkable instrument orbiting Earth since 1990. Named after the astronomer Edwin Hubble, who discovered that the universe is expanding, the Hubble Space Telescope has given us an unprecedented view of the cosmos.

The telescope is about the size of a school bus and orbits the Earth at about 340 miles. It is equipped with various instruments that allow it to observe objects in different wavelengths of light, from ultraviolet to infrared.

One of the most remarkable things about the Hubble Space Telescope is its ability to capture incredibly detailed images of objects in space. From distant galaxies to nebulae to stars and planets in our solar system, the Hubble Space Telescope has provided us with stunning views of the universe.

But the Hubble Space Telescope isn't just about pretty pictures. It has also made some of our most important scientific discoveries. For example, it has helped to refine our understanding of the age and size of the universe, as well as the processes that drive the formation of galaxies.

The telescope has also provided evidence for the existence of dark matter, a mysterious substance that makes up most of the matter in the universe. By observing how light is bent and distorted by gravity, astronomers have been able to map the distribution of dark matter in the universe.

Another important discovery made by the Hubble Space Telescope is the universe's accelerating expansion. This phenomenon, first observed by the telescope in 1998, suggests that the universe is filled with a mysterious force called dark energy, which is causing the expansion to accelerate.

The Hubble Space Telescope has also played a crucial role in searching for habitable planets outside our solar system. By observing how light from distant stars is dimmed as planets pass in front of them, astronomers have been able to detect hundreds of exoplanets, some of which may be capable of supporting life.

Despite its incredible accomplishments, the Hubble Space Telescope has faced its share of challenges. In 1990, shortly after it was deployed, astronomers realized that its primary mirror was flawed, causing images to be blurry. However, a team of engineers and astronauts were able to fix the problem by installing corrective optics during a series of spacewalks in 1993.

Since then, the Hubble Space Telescope has been serviced several times by space shuttle missions, with the most recent service mission taking place in 2009. However, with the space shuttle program's retirement in 2011, the Hubble Space Telescope is no longer being serviced by astronauts.

Despite this, the telescope still provides valuable scientific data and will continue to do so for years to come. Its successor, the James Webb Space Telescope, is set to launch in 2021 and will build on the incredible legacy of the Hubble Space Telescope, revealing even more secrets of the universe.

So there you have it, the story of the Hubble Space Telescope and how it has revealed the universe's secrets. It's an incredible achievement that has expanded our understanding of the cosmos and opened up new frontiers of scientific discovery.

Exploring the Outer Solar System
The Voyager Missions

In the late 1970s, NASA launched two spacecraft called Voyager 1 and Voyager 2 on a mission to explore the outer solar system. These missions would take the spacecraft past Jupiter, Saturn, Uranus, and Neptune, giving us an up-close look at these distant planets and their moons.

The Voyager spacecraft were about the size of a car and were equipped with various scientific instruments, including cameras, spectrometers, and magnetometers. They were also equipped with a golden record containing sounds and images intended to represent Earth to any intelligent extraterrestrial life that might encounter the spacecraft.

Voyager 2 was launched in August 1977, followed by Voyager 1 a few weeks later. The spacecraft followed different flight paths that would take them past other planets at different times, allowing scientists to maximize their observations.

The first destination for the Voyager spacecraft was Jupiter, which they reached in 1979. The spaceship made several close flybys of the planet, taking detailed images of its cloud patterns, moons, and other features. They also discovered that Jupiter had a complex system of rings, which had yet to be detected from Earth.

After leaving Jupiter, the Voyager spacecraft continued to Saturn, which they reached in 1980. Saturn was an even more spectacular sight than Jupiter, with iconic rings and dozens of moons. The

Voyager spacecraft took detailed images of the planet and its moons, revealing a wealth of new information about these distant worlds.

One of the most exciting discoveries made by the Voyager spacecraft was the presence of geysers on one of Saturn's moons, Enceladus. The spacecraft detected plumes of water vapour and ice particles erupting from the moon's surface, indicating the presence of a subsurface ocean.

After leaving Saturn, the Voyager spacecraft continued to Uranus and Neptune, which they reached in 1986 and 1989. These were the first and only spacecraft ever to visit these distant ice giants, and they provided us with our sole close-up views of these planets and their moons.

One of the most stunning discoveries made by the Voyager spacecraft was the presence of active geysers on Neptune's largest moon, Triton. The spacecraft detected plumes of nitrogen gas and dust erupting from the moon's surface, indicating it was still geologically active.

The Voyager missions were a remarkable achievement in space exploration, providing us with our first close-up views of the outer solar system. They also gave us a wealth of new scientific data about these distant worlds, helping us to understand their origins, geology, and atmosphere.

Today, the Voyager spacecraft are still flying through interstellar space beyond the outer reaches of our solar system. They continue to send back data, even after more than 40 years in space and are expected to continue doing so for many years to come.

Now you know the story of the Voyager missions and how they explored the outer solar system. It's an incredible achievement that expanded our understanding of the cosmos and opened up new frontiers of scientific discovery.

Uncovering Clues on Mars
The Curiosity Rover

In 2012, NASA launched a robotic rover called Curiosity to explore the surface of Mars. The Curiosity Rover was about the size of a car and was equipped with various scientific instruments, including cameras, spectrometers, and a laser that could vaporize rocks and soil for analysis.

The Curiosity mission aimed to investigate whether Mars had the right conditions to support life in the past and to study the planet's geology, climate, and environment.

Curiosity landed on Mars in August 2012, using a unique landing system that involved a parachute, retro-rockets, and a sky crane that lowered the rover onto the surface. The landing was a remarkable achievement, and the rover quickly began exploring its new surroundings.

One of the first things that Curiosity did was take a series of panoramic images of its landing site, Gale Crater. The photos showed a rocky, desert-like landscape with a large mountain called Mount Sharp in the distance.

Over the next few years, Curiosity explored the area around Gale Crater, driving over rocky terrain and analyzing the composition of the soil and rocks. It discovered that the site was once covered by water and that there were likely rivers and lakes in the past.

Curiosity also discovered organic molecules, which are the building blocks of life, in the rocks and soil on Mars. While these molecules don't necessarily indicate the presence of life on Mars, they do suggest that the planet had the right conditions to support life in the past.

One of the most exciting discoveries made by Curiosity was the presence of methane gas in the Martian atmosphere. Methane is a potential sign of biological activity, although non-biological processes can also produce it. The discovery of methane has renewed interest in the possibility of life on Mars.

Curiosity also studied the weather on Mars, measuring temperature, humidity, and wind speed. It discovered that the weather on Mars is much more extreme than on Earth, with temperatures that can swing from -200 degrees Fahrenheit at night to 70 degrees Fahrenheit during the day.

Throughout its mission, Curiosity has sent back thousands of images and scientific data, helping us better understand Mars's geology and environment. It has also paved the way for future Mars missions, including the upcoming Mars 2020 mission, which will launch a new rover to explore a different part of the planet.

So there you have it, the story of the Curiosity Rover and its mission to uncover clues on Mars. It's an incredible achievement that expanded our understanding of the Red Planet and brought us one step closer to answering the age-old question of whether life is beyond Earth.

Living in Space
The International Space Station

The International Space Station, or ISS for short, is a giant space laboratory that orbits Earth at an altitude of around 400 kilometers. It's the largest human-made object in space, and it's been continuously occupied by astronauts since 2000.

The ISS is a joint project between the United States, Russia, Europe, Japan, and Canada. It's about the size of a football field and weighs over 400,000 kilograms. The station comprises many different parts, including living quarters for the astronauts, laboratories for conducting experiments, and storage areas for supplies.

One of the most incredible things about the ISS is that it travels around the Earth at a speed of around 28,000 kilometers per hour, completing one orbit every 90 minutes. That means the astronauts onboard see a sunrise and sunset every 45 minutes!

Living in space is very different from living on Earth. For one thing, there's no gravity in space, so everything floats around. It can be a lot of fun, but it also takes some time to get used to it. Astronauts must be careful to strap themselves into their seats or use special hooks to keep from floating away.

Another challenge of living in space is that the human body reacts differently to the environment. For example, without gravity, the fluids in the body tend to float up to the head, causing astronauts to experience a puffy face and a feeling of congestion. That is why astronauts exercise several hours daily to help their bodies adjust to the environment.

Despite these challenges, the astronauts on the ISS carry out a wide range of scientific experiments in various fields, including physics, biology, and medicine. These experiments help us better understand the effects of living in space on the human body and develop technologies that can be used on Earth.

The astronauts on the ISS also have to be very self-sufficient. They grow their food using special hydroponic systems and recycle their waste products to produce clean water for drinking and other uses. The station also has special air filters to keep the air clean and breathable.

Living on the ISS can be a lot of fun too. The astronauts have a great view of the Earth and can see amazing sights like the Northern Lights and shooting stars. They can also participate in special activities like spacewalks, where they venture outside the station to conduct repairs or experiments.

To bring to a close, the International Space Station is an incredible feat of human engineering and an amazing place to live and work. The astronauts onboard face many challenges, but they also have the opportunity to carry out groundbreaking scientific research and to experience the wonder of space firsthand. It reminds us of the incredible things that can be achieved when people work together to explore and discover the world around us.

The Dark Secrets of the Universe
Black Holes

Black holes are one of the universe's most mysterious and fascinating objects. They are areas in space where the gravitational pull is so strong that nothing, not even light, can escape. That makes them invisible to the naked eye and difficult to study.

Black holes form when massive stars run out of fuel and collapse under their weight, creating a region where the gravitational pull is so strong that it warps the fabric of space and time. This region is called the event horizon, and anything that crosses it is sucked into the black hole, never to return.

One of the most interesting things about black holes is that they affect the objects around them in strange ways. For example, if a star gets too close to a black hole, it can be torn apart by gravitational forces. It creates a bright flare of radiation that telescopes on Earth can detect.

Another interesting phenomenon associated with black holes is the creation of powerful jets of gas and radiation that shoot out from the black hole at nearly the speed of light. These jets can travel millions of light-years across space and can be detected by radio telescopes.

Despite their invisible nature, astronomers have been able to study black holes using various techniques. One method is to observe the effects of their gravity on nearby stars and gas clouds. By measuring the movements of these objects, astronomers can infer the presence of a black hole.

Another method is to observe the radiation emitted by material falling into a black hole. This radiation can give clues about the properties of the black hole, such as its mass and spin.

Black holes are not just fascinating objects for study but also play an important role in the universe. For example, they are thought to be responsible for powering the quasars, the most luminous objects in the universe. Quasars are galaxies that contain supermassive black holes at their centers, which are surrounded by disks of gas and dust. As the material falls into the black hole, it heats up and emits enormous amounts of radiation.

Black holes also play a role in the evolution of galaxies. As galaxies merge, their black holes can join as well, creating even larger black holes that can profoundly affect the structure and dynamics of the universe.

So as you can understand now, black holes are mysterious and fascinating objects in the universe that have captured the imaginations of scientists and the public. Although they are invisible and difficult to study, astronomers have learned a great deal about them using various techniques. Black holes play important roles in the universe, from powering quasars to shaping the structure of galaxies. They are a reminder of the incredible mysteries that still exist in the cosmos and the endless possibilities for discovery and exploration.

The Beginning of Everything
The Big Bang

The Big Bang is the scientific explanation for how the universe began. Scientists believe it all started about 13.8 billion years ago when the universe was just a tiny, infinitely dense, hot point called a singularity. Then, all of a sudden, it expanded rapidly in a massive explosion that we call the Big Bang.

Now, you might be wondering how scientists know all of this. Well, they have used telescopes to study the light from faraway galaxies, and they have found that the light is shifted toward the red end of the spectrum. It is called redshift, which tells us that those galaxies are moving away from us. If you imagine rewinding time and bringing those galaxies closer and closer together, you eventually get to a point where they are all in one place. And that's what the universe was like at the beginning of the Big Bang.

So, the universe continued to expand and cool down after the Big Bang. As it cooled, matter began to form, including atoms and molecules. These eventually came together to form stars and galaxies.

The Big Bang also left behind a lot of energy in the form of radiation. This radiation, called the cosmic microwave background, is still present today, and it is one of the pieces of evidence supporting the Big Bang theory.

Now, it's important to remember that the Big Bang wasn't an explosion, as we might think. There was no center of the blast and no outside to the universe for it to expand into. Instead, the entire universe expanded simultaneously and in all directions, like a balloon inflating.

So, that's the story of the Big Bang. It's a unique and important discovery that helps us understand how the universe began and evolved over billions of years.

Powerful Eruptions from the Sun
Solar Flares

The sun is a vast, hot ball of gas that is constantly producing energy through a process called nuclear fusion. Sometimes, this energy builds up and is released in a burst of activity called a solar flare.

Solar flares can be very powerful and release as much energy as millions of nuclear bombs. They usually occur near sunspots, areas on the sun's surface that are cooler than the surrounding area. Sunspots are caused by the sun's magnetic field, which can become twisted and knotted in these regions.

When a solar flare occurs, it releases a burst of high-energy particles, including X-rays and gamma rays. These particles can travel through space and can even affect the Earth. If a solar flare is aimed toward the Earth, it can cause a geomagnetic storm, disrupting our technology, including satellites and power grids.

But solar flares can also create some beautiful sights in the sky. When the high-energy particles from a solar flare interact with the Earth's magnetic field, they can cause a phenomenon called the aurora borealis, or northern lights. These colorful, shimmering lights appear in the sky near the Earth's poles.

Scientists study solar flares to understand better the sun and how it affects our planet. They use telescopes to observe the sun and track the activity of sunspots and solar flares. This information can help us predict when a solar flare might occur and how it might affect us.

So, that's the story of solar flares. They are powerful eruptions from the sun that can affect the Earth and create beautiful sights in the sky. While they can be dangerous, they also help us learn more about the sun and our place in the universe.

A Dazzling Display in the Sky
The Northern Lights

The Northern Lights, also known as Aurora Borealis, is one of the most spectacular displays in the sky. It is a natural light show that occurs in the Earth's polar regions. The Northern Lights appear as a colorful display of light in the sky, with colors ranging from green to pink, purple, and even red.

The Northern Lights are created by charged particles ejected from the Sun. These particles travel toward the Earth and are attracted by our planet's magnetic field. They make a beautiful display of light as they collide with the gasses in the Earth's atmosphere.

The Northern Lights can be seen in countries such as Norway, Sweden, Finland, Canada, and Iceland. To witness this breathtaking display, one must travel to a location with a clear view of the northern sky and away from any light pollution.

The indigenous people of the Arctic, such as the Inuit and the Sami, have their folklore and legends surrounding the Northern Lights. The Inuit believed that the lights were the spirits of their ancestors playing a game of soccer with a walrus skull. The Sami thought that the lights were the souls of their ancestors dancing in the sky.

Scientists have also studied the Northern Lights to learn more about our planet's atmosphere. By analyzing the light emitted during an aurora, scientists can determine the chemical makeup of the gasses in the atmosphere.

In addition, the Northern Lights can also have an impact on our technology. The charged particles that create the Northern Lights can interfere with radio and satellite communications and power grids. That is why scientists and engineers study the Northern Lights to understand better how they can affect our technology.

The Northern Lights are indeed a spectacular sight to behold. They remind us of the beauty and mystery of our planet and the universe. By studying them, we can learn more about our world and the forces that shape it.

In conclusion, the Northern Lights are a dazzling display of nature's beauty that has captivated people for centuries. They are created by charged particles from the Sun colliding with gases in the Earth's atmosphere, resulting in a colorful display of light. While the Northern Lights are certainly a sight to behold, they also have scientific and cultural significance. Studying them can help us better understand our planet's atmosphere and the impact of the Sun's activity on our technology.

The Region Beyond Neptune
The Kuiper Belt

The Kuiper Belt is a region of our Solar System that is located beyond the planet Neptune. It is named after astronomer Gerard Kuiper, who proposed its existence in 1951. The Kuiper Belt is a vast, doughnut-shaped area home to many small icy bodies, including dwarf planets, comets, and asteroids.

Scientists believe that the Kuiper Belt is made up of the leftover materials from the formation of the Solar System, which happened about 4.6 billion years ago. These materials never coalesced into a planet like the ones we have today. Instead, they formed into smaller icy objects that have been orbiting the Sun ever since.

One of the most famous objects in the Kuiper Belt is the dwarf planet Pluto, discovered in 1930 by astronomer Clyde Tombaugh. Pluto was considered the ninth planet of our Solar System until 2006, when the International Astronomical Union reclassified it as a dwarf planet. Since then, dwarf planets have been discovered in the Kuiper Belt, including Eris, Haumea, and Makemake.

The Kuiper Belt is difficult to study because it is so far from Earth. Spacecraft take a long time to travel there, and the objects in the Kuiper Belt are tiny and faint. In 2015, NASA's New Horizons spacecraft made a historic flyby of Pluto, giving us our first close-up look at this distant world. New Horizons then continued into the Kuiper Belt, studying a small object called Arrokoth in 2019.

Scientists are still learning a lot about the Kuiper Belt and its objects. By studying them, they hope to understand better the early Solar System and how it formed. Some scientists also believe that the Kuiper Belt could be a source of resources for future space exploration, as it contains a lot of water, ice, and other valuable materials.

The Kuiper Belt is a fascinating region of our Solar System, home to many small icy objects. It holds important clues to the formation of our Solar System and could be a valuable resource for future space exploration. Although it is far away and challenging to study, scientists continue to make important discoveries about this distant and mysterious region.

The Edge of the Solar System
The Oort Cloud

The Solar System is a vast expanse of space home to our Sun, eight planets, numerous asteroids, and comets. Beyond the orbit of Neptune, the outermost planet in the Solar System lies a mysterious region called the Oort Cloud.

The Oort Cloud is named after the Dutch astronomer Jan Oort, who proposed its existence in 1950. This region is considered a vast, spherical cloud of icy objects that encircles the Solar System and stretches out to a distance of up to 100,000 astronomical units (AU) from the Sun. For context, one AU is the average distance between the Earth and the Sun, about 93 million miles.

Scientists believe the Oort Cloud is the source of long-period comets that occasionally visit the inner Solar System. These comets take thousands of years to orbit the Sun and originate from the distant reaches of the Oort Cloud.

Unlike the Kuiper Belt, another region beyond Neptune, the Oort Cloud is believed to be a more remote and sparse region. Studying is also much more complicated because no spacecraft has ever visited it. However, scientists can learn about the Oort Cloud by studying comets that originate from there.

Comets are icy bodies that are thought to be remnants of the early Solar System. As a comet approaches the Sun, its icy surface vaporizes, creating a bright tail that can be seen from Earth. By studying the composition of comets, scientists can learn about the conditions in the early Solar System.

The Oort Cloud is also thought to contain many undiscovered dwarf planets, which are more petite than planets but more significant than asteroids. These dwarf planets would be challenging to detect because of their distance from the Sun and the lack of sunlight that reaches them.

While we still have much to learn about the Oort Cloud, it is an important region that helps us understand the formation and evolution of our Solar System. By studying the comets and icy objects that originate from the Oort Cloud, scientists can learn about the conditions in the early Solar System and how our Solar System came to be.

In summary, the Oort Cloud is a vast, mysterious region at the Solar System's edge. It is believed to be a source of long-period comets and may contain undiscovered dwarf planets. While it is a challenging region to study, it provides valuable insights into the formation and evolution of our Solar System.

Jupiter's Famous Storm
The Great Red Spot

Have you ever looked at the night sky and wondered what lies beyond our planet? One of the fascinating things in our solar system is Jupiter, the giant planet. Jupiter is so big that all the other planets in our solar system could fit inside it. And one of the most exciting features of Jupiter is the Great Red Spot, a giant storm that has been raging for hundreds of years.

The Great Red Spot is a massive storm that swirls around Jupiter's atmosphere. It's so huge that it's even more significant than the size of Earth! The storm is believed to have been first observed by humans in the 17th century, making it one of the oldest known storms in our solar system. And scientists estimate that it has been raging for over 300 years!

The storm is called the Great Red Spot because of its color. It's a deep, brick-red color that makes it stand out from the rest of Jupiter's atmosphere. The color is believed to be caused by chemicals in the storm's atmosphere, such as ammonia and methane, that Jupiter's strong winds affect.

Scientists have been studying the Great Red Spot for decades, using powerful telescopes and even spacecraft to gather data about the storm. They have discovered that the storm is not only massive but also compelling. The winds inside the storm can reach speeds of up to 400 miles per hour! That's more than twice as fast as the strongest winds ever recorded on Earth.

Despite its size and power, the Great Red Spot has been slowly shrinking. In the late 1800s, it was estimated to be about four times the size of Earth. But today, it's only about 1.3 times the size of our planet. Scientists are still trying to figure out why the storm is shrinking and what that could mean for Jupiter's atmosphere.

But one thing is sure: the Great Red Spot is a spectacular sight in our solar system. It's a reminder that there is so much we still have to learn about the universe and the amazing things that can happen in space. And who knows, maybe one day in the future, humans will be able to visit Jupiter and see the Great Red Spot up close.

Discovering New Worlds
Exoplanets

Exoplanets are planets that orbit stars other than our Sun. Astronomers have discovered thousands of exoplanets since the first one was found in 1995, and many more are expected to be found in the coming years. It is an exciting field of study because exoplanets could be home to life like Earth.

We need to know how planets are detected to understand how scientists find exoplanets. Several methods are used to discover exoplanets, but the most common is the transit method. That involves observing a star and looking for any dips in its brightness caused by a planet passing in front of it. By studying the timing and depth of these dips, scientists can determine the size, orbit, and even some of the planet's properties.

Another method is the radial velocity method, which involves observing a star's movement as it is gravitationally pulled by the planet orbiting around it. By measuring the star's wobble, scientists can determine the planet's mass and orbit.

Once exoplanets are discovered, scientists can use telescopes to study their atmospheres and determine if they have the right conditions for life to exist. For example, they look for signs of water vapor, oxygen, and other gases typically associated with life on Earth.

One of the most exciting discoveries in the field of exoplanets was the discovery of Proxima b in 2016. It is an Earth-sized planet that orbits in the habitable zone of Proxima Centauri, the closest star to our Sun. There could be other planets in this system that could also support life.

As technology advances, scientists can discover smaller and more distant exoplanets and even study their atmospheres in more detail. There is still much to learn about these fascinating worlds, and it is an exciting time to be studying exoplanets.

In conclusion, exoplanets are planets that orbit stars other than our Sun, and scientists have discovered thousands of them using various methods. They are an exciting field of study because they could be home to life, and scientists are studying their atmospheres to learn more about their potential habitability.

A Collision Course in Space
The Asteroid Belt

Once upon a time, a group of objects traveled through space between Mars and Jupiter. This group of things was called The Asteroid Belt, containing many rocks, ice, and dust. The Asteroid Belt formed around 4.6 billion years ago when our solar system was created.

The Asteroid Belt is a vast region about as wide as the distance between the Earth and the Sun. It is made up of millions of asteroids of different sizes. Some asteroids are tiny, while others can be several hundred kilometers across. The giant asteroid in the belt is called Ceres, also classified as a dwarf planet.

The asteroids in the belt need to be evenly spaced out. Some areas have more asteroids than others. Scientists think that this might be because of the gravitational pull of Jupiter. Jupiter is the giant planet in our solar system, and its gravity can affect the movement of objects in space.

Asteroids are made up of different materials. Some asteroids are made up of rock and metal, while others are made up of ice and dust. Scientists believe some asteroids have water and other vital resources that could be useful for future space exploration.

Asteroids move in a variety of ways. Some asteroids are on a collision course with planets, while others are on stable orbits around the Sun. Sometimes asteroids collide, causing fragments to break off and form more minor asteroids.

The Asteroid Belt is also home to many smaller objects called meteoroids. Meteoroids are tiny pieces of space debris that move through space. When a meteoroid enters the Earth's atmosphere, it creates a bright streak of light called a meteor or shooting star.

Although the idea of asteroids colliding with Earth might seem scary, scientists have been studying The Asteroid Belt to understand its behavior and potential threats. They have discovered that very few asteroids pose a significant threat to Earth.

Some asteroids could be helpful in space exploration. NASA has sent missions to study asteroids up close, including the OSIRIS-REx mission to asteroid Bennu. Scientists hope analyzing asteroids could help us learn more about the early solar system and the origins of life on Earth.

In conclusion, The Asteroid Belt is a fascinating region in our solar system full of mysteries and potential discoveries. While some asteroids could threaten Earth, scientists are actively studying the belt to understand its behavior and learn more about our solar system's history. Who knows what we might discover next in The Asteroid Belt?

Our Home Galaxy
The Milky Way

Let's talk about something extraordinary - our home, the Milky Way galaxy.

The Milky Way is a giant, beautiful galaxy filled with stars, planets, and many other cool things. But before discussing that, let's first understand what a universe is.

A galaxy is a massive collection of stars, planets, and other objects held together by gravity. There are billions of galaxies in the universe, each unique in its own way.

Our galaxy, the Milky Way, is shaped like a big spiral. Imagine a giant pinwheel, but instead of colors, it comprises stars and space objects. It is so big that if you were to travel at the speed of light (which is the fastest speed possible), it would take you over 100,000 years to travel from one end of the Milky Way to the other!

The Milky Way has a supermassive black hole in the center that is millions of times more massive than our sun. It is called Sagittarius A* and is believed to be over 4 million times more massive than the sun.

But the Milky Way is not just made up of stars and black holes. There are also other space objects, like nebulae and clusters, which are made up of gas and dust. Their bright colors and patterns are so beautiful to look at.

Now, let's talk about the stars. The Milky Way has billions of stars of different sizes and colors. Our sun is just one of those stars and pretty tiny. Some stars are much bigger than our sun, and others are smaller.

Did you know that there are planets in the Milky Way too? Our solar system is just one of the many in the Milky Way. Scientists have discovered thousands of planets outside our solar system, called exoplanets. Some of these exoplanets may even be able to support life!

So, the Milky Way is a pretty fantastic place. It is our home, and we are lucky to be a part of it. Next time you look up at the night sky and see the stars, think about how incredible it is that we are all a part of this vast, beautiful galaxy.

Explosions in the Sky
Supernovas

Have you ever looked up at the night sky and seen a bright star suddenly appear out of nowhere, only to disappear just as quickly? You may have witnessed a supernova, one of the universe's most significant and spectacular events.

A supernova is a massive explosion that occurs when a star runs out of fuel and collapses. The resulting blast can be so bright that it briefly outshines an entire galaxy. A supernova can release as much energy as the sun will produce in its lifetime!

There are two types of supernovas: Type I and Type II. Type I supernovas occur when a white dwarf star, the remnant of a smaller star like our sun, absorbs too much material from a nearby companion star. It causes the white dwarf to exceed a critical mass, collapsing under its gravity and resulting in a massive explosion.

Type II supernovas occur when a much larger star, at least eight times the mass of our sun, runs out of fuel and can no longer produce the energy necessary to support its weight. The star then collapses, and the resulting explosion can be so bright that it can be visible from Earth even during the daytime.

Supernovas are essential because of their spectacular display and because they are responsible for creating many of the heavy elements that make up our universe. Elements like gold, silver, and uranium are created during a supernova's explosion and are scattered throughout the galaxy, eventually forming new stars and planets.

While supernovas are incredibly powerful and can be seen from millions of light-years away, they are also rare. On average, only one supernova occurs in our galaxy every 50 years. So, if you're lucky enough to witness one in your lifetime, consider yourself very fortunate!

So next time you look up at the night sky, remember that those twinkling stars may have some explosive secrets. Supernovas may be rare, but they are an awe-inspiring reminder of the incredible power and beauty of the universe.

The Space Race
The US and USSR Compete for Space Supremacy

The Space Race was a competition between the United States and the Soviet Union to see who could be the first to achieve significant milestones in space exploration. It began in the late 1950s and lasted until the early 1970s. Both countries were determined to prove their technological and military superiority.

The Space Race officially began on October 4, 1957, when the Soviet Union launched Sputnik 1, the first artificial satellite, into space. This event caused shock and concern in the United States, as it showed that the Soviet Union could launch missiles that could reach America. The US responded by launching its first satellite, Explorer 1, four months later.

The Space Race intensified in 1961 when Soviet cosmonaut Yuri Gagarin became the first human to orbit the Earth. A few weeks later, American astronaut Alan Shepard became the first American to reach space, although he only made a brief suborbital flight.

In 1969, the United States achieved one of the most significant milestones of the Space Race by landing astronauts on the moon. The Apollo 11 mission, commanded by Neil Armstrong, marked the first time humans had set foot on another celestial body. This achievement was the result of years of hard work and technological advances.

The Space Race continued throughout the 1970s, with both countries launching space stations and conducting experiments in orbit. In 1975, the United States and the Soviet Union conducted their first joint space mission, the Apollo-Soyuz Test Project, which marked the end of the Space Race.

The Space Race was a critical moment in human history. It showed that humans could achieve incredible things when they work together and push possible boundaries. It also helped to advance technology and science, leading to many of the modern conveniences we enjoy today.

So, that was the story of the Space Race! I hope you enjoyed learning about it. Remember, the sky is not the limit when exploring and understanding the universe around us.

A Stellar Nursery
The Orion Nebula

Have you ever looked up at the night sky and seen a group of stars that seem to be huddled together in a shape that looks like a person or an animal? Those are called constellations; humans have used them for thousands of years to navigate and tell stories.

One constellation you can see during winter is Orion the Hunter. Orion comprises some of the brightest stars in the sky and is easily recognizable by his belt of three stars in a row. If you look closely at Orion's sword, you might notice a fuzzy patch of light. That fuzzy patch is the Orion Nebula, one of the most famous and fascinating objects in the night sky.

The Orion Nebula is located about 1,344 light-years away from Earth, meaning that the light we see from it now left the nebula over a thousand years ago! The nebula is a giant cloud of gas and dust that spans about 24 light-years across. That's more than 1,400 times the size of our solar system!

The Orion Nebula is so interesting because it's a stellar nursery. Inside the cloud, there are thousands of baby stars being born. Astronomers have been studying the Orion Nebula for many years, trying to understand how stars are born and how they evolve. They use telescopes to analyze the light and radiation coming from the nebula, which can tell them a lot about what's happening inside.

One thing that astronomers have discovered is that the Orion Nebula is a very active place. There are intense magnetic fields, powerful winds, and even jets of gas shooting out from some of the baby stars. All of this activity helps to shape the nebula and create new leads.

But why is the Orion Nebula so important to us here on Earth? Well, stars like our sun are responsible for creating many elements that make up our bodies and the world around us. The carbon in our cells, the oxygen we breathe, and even the calcium in our bones were all created inside stars. By studying star formation in the Orion Nebula, astronomers can learn more about the origins of the elements that make up our world.

So the next time you see Orion the Hunter in the night sky, remember that his sword contains a giant nursery where stars are born. And who knows, someday one of those baby stars will have planets like ours, and maybe even life!

The Red Planet
Exploring Mars

Let's talk about a planet that has always fascinated people for centuries: Mars, also known as the Red Planet.

Mars is the fourth planet from the Sun, named after the Roman god of war. It earned this nickname because of the reddish color on its surface, which comes from iron oxide, also known as rust. Mars is a rocky, terrestrial planet with a thin atmosphere, and it's also the second closest planet to Earth after Venus.

Mars has been a popular subject of exploration for many years. Scientists have been trying to study and explore Mars for centuries. And in recent times, various missions have been sent to Mars to learn more about its atmosphere and geology and search for evidence of past or present life.

One of the most significant explorations of Mars was NASA's Viking missions in the 1970s, which were the first to land spacecraft on Mars. These missions provided us with images and data that helped us understand more about Mars, such as its thin atmosphere and the presence of water in the form of ice.

In recent years, NASA's Mars rovers, Curiosity and Perseverance have continued to explore the Red Planet's surface. Curiosity landed on Mars in 2012 and has been exploring the Gale Crater, a massive impact crater on Mars. It has been studying the rocks and soil to learn more about the planet's geological history and whether it had the necessary ingredients to support life in the past.

On the other hand, Perseverance landed on Mars in February 2021 and is on a mission to search for signs of ancient microbial life. It will collect samples of the Martian soil and rocks, which will be returned to Earth for further analysis. The samples will help us understand the history of Mars and whether it could have supported life at some point.

Apart from NASA, other countries such as Russia, China, and the United Arab Emirates have also sent missions to Mars in recent years. With all these missions, we are learning more and more about Mars and its potential for life.

Who knows, someday, humans will even be able to visit Mars and explore it firsthand. We're still a long way from that, but with continued exploration and research, we can unlock the secrets of Mars and the universe beyond.

Mystery in Orbit
The Black Knight Satellite

Have you ever heard of the Black Knight Satellite? It's a mysterious object in space that has fascinated people for years. It's a satellite that is said to be orbiting the Earth, but no one knows for sure what it is or where it came from.

The story of the Black Knight Satellite began in the 1950s when some strange objects were detected in space. Some people thought these objects might be satellites, but at that time, no one had launched satellites into space. Some even believed that the things might be alien spacecraft.

Then, in 1960, a mysterious object was detected by radar. It was orbiting the Earth in a polar orbit, which is unusual for satellites. The thing was dubbed the Black Knight Satellite and has been the subject of much speculation ever since.

Some people believe that the Black Knight Satellite is an alien spacecraft observing Earth for centuries. Others think it might be a secret military satellite launched during the Cold War. And still, others believe it is simply a piece of space debris that has been mistaken for a satellite.

Despite all the speculation, we still don't know exactly what the Black Knight Satellite is. Some have tried to study it with telescopes, but it isn't easy to get a good look at something so far away and to move so quickly.

But even if we don't know the Black Knight Satellite precisely, it's still an exciting mystery that captures our imaginations. Who knows? One day we'll discover the truth behind this mysterious object in space.

Our Closest Neighbor
The Andromeda Galaxy

Did you know there are billions of galaxies in the universe, each with millions or billions of stars? Our galaxy, the Milky Way, is just one of them. But did you also know that our closest neighbor galaxy is Andromeda? That's right; Andromeda is only about 2.5 million light-years away from us. That may sound far, but it's close in space terms!

Andromeda is also known as M31 because it is the 31st object in a catalog of celestial objects created by astronomer Charles Messier. It is named after the constellation Andromeda, which is named after a character in Greek mythology.

Andromeda is a spiral galaxy, just like the Milky Way. It has a central bulge of stars and a disk of gas, dust, and stars that spiral outward from the center. Two smaller companion galaxies, M32 and M110, are part of the Andromeda family.

Scientists think Andromeda has about 1 trillion stars, about twice as many as the Milky Way. It is also more significant than the Milky Way, with a diameter of about 220,000 light-years.

One of the reasons why Andromeda is so enjoyable to scientists is because it is a "starburst" galaxy. It means that it is currently forming new stars at a much higher rate than the Milky Way. It is estimated that Andromeda produces about 1 solar mass of new stars per year, about 5 times more than the Milky Way.

But don't worry; Andromeda will not collide with the Milky Way anytime soon! Scientists estimate that the two galaxies will merge in about 4.5 billion years. That may sound like a long time, but it's just a blink of an eye in cosmic terms.

Even though Andromeda is our closest neighbor galaxy, it can be hard to see with the naked eye. It would help if you were in a place with very dark skies, away from city lights. But if you can see Andromeda through a telescope, it's a fantastic sight. You'll look at the light that left Andromeda over 2 million years ago!

So, there you have it, the Andromeda Galaxy, our closest neighbor in space. It's a beautiful spiral galaxy, much like our Milky Way, but with interesting differences. Who knows what else we'll discover about Andromeda in the future?

The Result of a Stellar Explosion
The Crab Nebula

Let's talk about one of the most amazing things in our universe - the Crab Nebula! It's a beautiful object in the sky formed from the remnants of a supernova explosion.

But before we get into the Crab Nebula itself, let's talk briefly about what a supernova is. A supernova is an explosion when a star runs out of fuel and can no longer support its weight. When this happens, the star's core collapses and then explodes outward in a burst of energy. This explosion can be so bright that it can briefly outshine an entire galaxy!

Now, back to the Crab Nebula. It's located in the constellation Taurus and is one of the most studied objects in the night sky. The Crab Nebula was created by a supernova explosion observed by Chinese and Japanese astronomers in 1054. It was so bright that it was visible during the day for several weeks!

The explosion created a rapidly expanding cloud of gas and dust called the Crab Nebula. It's about 6,500 light-years away from Earth, which means the light we see from it today left the nebula 6,500 years ago.

The Crab Nebula is an incredibly complex object, with a central pulsar and a tangled web of filaments and knots of gas and dust. The pulsar is the remnant of the star that exploded and is spinning at an incredible rate - about 30 times per second! It also emits radiation beams that sweep around like a lighthouse, which is why we call it a pulsar.

Scientists have been studying the Crab Nebula for decades, and they're still learning new things about it all the time. They've used telescopes and other instruments to study the nebula in different wavelengths of light, from radio waves to X-rays. They've also sent spacecraft to study it up close, including the Hubble Space Telescope and the Chandra X-ray Observatory.

One of the things that scientists have learned from studying the Crab Nebula is that it's not just a beautiful object in the sky - it's also a laboratory for studying some of the most extreme physics in the universe. For example, the Crab Nebula is one of the brightest sources of gamma-ray radiation in the sky, produced when high-energy particles are accelerated to nearly the speed of light.

In conclusion, the Crab Nebula is an incredible object in the sky formed from the remnants of a supernova explosion. It's located in the constellation Taurus, about 6,500 light-years away from Earth, and is one of the most studied objects in the night sky. Scientists have been studying the nebula for decades and have used telescopes and spacecraft to learn more about its complex structure and the extreme physics that occur within it.

Searching for Extraterrestrial Life
The Drake Equation

Have you ever wondered if other intelligent beings are living in the vast expanse of the universe? Well, many scientists and astronomers have asked this question too, and they have devised a special formula to help them estimate the likelihood of finding extraterrestrial life. This formula is known as the Drake equation, named after the astronomer Frank Drake who developed it in 1961.

The Drake equation is a mathematical equation that estimates the number of communicative civilizations that may exist in our galaxy. It takes into account several factors that scientists believe are important in the development of life, such as the number of stars in the galaxy, the fraction of stars that have planets, the fraction of habitable planets, the fraction of habitable planets that develop life, and the fraction of those that go on to develop intelligent life capable of communicating with us.

Now, let's break down the Drake equation a little further to understand better what it means. The equation's first factor is the star formation rate in our galaxy. Scientists estimate that our Milky Way galaxy has between 100 and 400 billion stars. The next factor is the fraction of stars that have planets. Recent discoveries suggest that most stars have at least one planet orbiting around them.

The third factor in the Drake equation is the number of planets in their star's habitable zone. It is the region around a star where conditions are just right for liquid water to exist on the surface of a planet, a key ingredient for life as we know it. The fourth factor is the fraction of habitable planets that develop life. It is the most uncertain factor in the equation, as we have no direct evidence of life anywhere other than Earth.

The fifth and final factor is the fraction of civilizations that develop a technology that enables them to communicate across space. This factor is also difficult to estimate, as we don't know how common or rare intelligent life is in the universe.

By combining all these factors in the Drake equation, we can estimate the number of intelligent, communicative civilizations in our galaxy that we can detect or communicate with. However, it's important to note that this is only an estimate, and the actual number could be much higher or lower.

So far, we haven't found any definitive evidence of extraterrestrial life, but that doesn't mean it's not out there. Scientists continue to search the skies for signs of life beyond Earth, using tools like radio telescopes and spacecraft to explore our solar system and beyond. Who knows what discoveries await us in the future?

Seven Earth-Like Planets?
The Trappist-1 System

Have you ever wondered if there are other planets out there like Earth? Well, scientists have been asking this question for a long time too! In 2017, astronomers discovered a unique star system called Trappist-1, which has seven planets that are all similar in size to Earth!

Trappist-1 is a tiny, dim star about 40 light-years away from us. Its seven planets are so close to the star that they orbit it in just a few days! Because they are so close to their star, scientists believe some planets may have water on their surface, just like Earth. Water is important for life as we know it, so this makes Trappist-1 an incredibly exciting discovery.

The planets in the Trappist-1 system have been given letter names, like Trappist-1b, Trappist-1c, etc. They are all very close together, so if you were standing on one of the planets, you would see the others as giant orbs in the sky! Some of the planets are so close that if you were standing on one of them, you could see the colors of the other planets and even the other planet's oceans!

Scientists are still studying the Trappist-1 system to learn more about these fascinating planets. They want to know if they have atmospheres if they have oceans, and if there might be life on any of them. They use powerful telescopes to study the light coming from the Trappist-1 star and its planets to find out. They can learn a lot about a planet from the light it reflects or the light it blocks when it passes in front of its star.

Trappist-1 is a very exciting discovery because it shows us that there are other star systems with planets similar to Earth. Who knows, maybe someday we will find life on one of these planets! Until then, scientists will keep studying and exploring the mysteries of the universe.

The Oldest Light in the Universe
The Cosmic Microwave Background

The story of the cosmic microwave background begins with the Big Bang, the event that marked the beginning of our universe over 13 billion years ago. According to the Big Bang theory, the universe started as an incredibly hot and dense point that expanded rapidly and cooled down over time. As it cooled, particles such as protons and electrons began to come together to form atoms, which allowed light to travel more freely through space.

Let's fast forward to about 380,000 years after the Big Bang. At this point, the universe had cooled down enough that the first atoms could form. This event is known as recombination, and it was a crucial turning point in the universe's history. Before recombination, the universe was filled with a hot, dense plasma that was opaque to light, meaning that photons (particles of light) couldn't travel very far before they collided with an electron or proton and were scattered in a new direction. But once the first atoms formed, the universe suddenly became transparent, and photons could travel freely through space.

The light released at the time of recombination has been traveling through the universe ever since, gradually getting stretched out and becoming less energetic as the universe has continued to expand. Today, this light appears to us as a faint glow visible in all directions of the sky. This glow is the cosmic microwave background.

In the 1960s, two astronomers, Arno Penzias, and Robert Wilson, were using a radio telescope to study the universe when they discovered a faint, constant hiss of microwave radiation coming from every direction in the sky. At first, they thought the hiss was just a result of some technical glitch in their equipment. But after ruling out all other possible sources of interference, they realized they were seeing the cosmic microwave background.

The discovery of the cosmic microwave background was a breakthrough in our understanding of the universe. It provided strong evidence for the Big Bang theory and helped us map out the universe's large-scale structure. By studying the patterns and fluctuations in the cosmic microwave background, astronomers have learned a great deal about the distribution of matter and energy in the early universe, as well as the properties of dark matter and dark energy.

Today, astronomers continue to study the cosmic microwave background using ever-more sophisticated telescopes and instruments. By analyzing the temperature and polarization patterns in the CMB, they hope to gain new insights into the nature of the universe and the fundamental physics that governs it. Who knows what discoveries await us in the future as we continue to explore the oldest light in the universe?

A Celestial Eye
The Helix Nebula

Have you ever looked up at the night sky and wondered about the many beautiful objects you can see twinkling above you? One of the most amazing sights in the universe is the Helix Nebula, also known as the Eye of God or the Eye of Sauron.

The Helix Nebula is a large, beautiful object in the sky that looks like an eye. It is located in the constellation Aquarius, which means it can be seen from the Earth's southern hemisphere. The nebula is around 700 light-years away from us, which is quite far but still visible to the naked eye.

The Helix Nebula is a type of nebula known as a planetary nebula. These nebulae are formed when a star similar to our Sun runs out of fuel and starts to die. As the star's core collapses, it sends a shockwave that pushes the outer layers of gas and dust into space. The gas and dust are illuminated by the dying star's intense ultraviolet light, creating the beautiful and colorful glowing cloud we see as a planetary nebula.

The Helix Nebula is one of the closest planetary nebulae to Earth, making it one of the most studied objects of its kind. It has a distinctive shape, with two concentric rings of gas and dust that appear to be swirling around a bright central star. The outer ring of the nebula is about 2.5 light-years across, while the inner ring is about 0.8 light-years across. The central star of the nebula is a white dwarf, which is the remnant of the star that created the nebula.

Scientists study the Helix Nebula to learn more about how stars evolve and die and also to learn more about the structure of our galaxy. By studying the light emitted by the nebula, scientists can determine the chemical composition of the gas and dust within it. They can also use the nebula to study the motion of stars in our galaxy and the distance to other galaxies in the universe.

The Helix Nebula is an incredible object in the night sky that reminds us of the amazing beauty and complexity of our universe. It is a celestial eye that has been gazing down at us for millions of years and will continue to do so for millions more.

A Cosmic Mystery
The Great Attractor

Now, we will talk about a fascinating mystery in the universe - the Great Attractor. The Great Attractor is a region of space pulling everything around it towards itself, including our Milky Way galaxy. Scientists have been studying the Great Attractor for many years, trying to unravel the secrets of this mysterious force.

The Great Attractor is located in the direction of the constellation Centaurus, about 200 million light-years away from us. It was first discovered in the 1970s when astronomers noticed that galaxies in that direction were moving towards a common point in space instead of being randomly distributed. This observation led them to believe there must be some kind of massive object in that region of space, causing the galaxies to move in a particular direction.

The Great Attractor is not visible to the naked eye or with telescopes that use visible light. It can only be detected by studying the motions of galaxies in that direction. Scientists believe that the Great Attractor is made up of a vast cluster of galaxies, which is so massive that it exerts a gravitational force on everything around it, causing them to move towards it.

One of the challenges in studying the Great Attractor is that it lies in a region of space with a lot of dust and gas, which can obscure our view. To get a clearer picture of what is happening in that space region, scientists have to use other wavelengths of light, such as infrared and radio waves, which can penetrate through dust and gas.

Despite many years of study, we still don't know much about the Great Attractor. Scientists are still trying to figure out exactly what is causing the gravitational pull in that space region and what objects are present there. Some have even speculated that the Great Attractor could be a supermassive black hole or an exotic object we haven't yet discovered.

In conclusion, the Great Attractor is a fascinating cosmic mystery scientists are still trying to unravel. It is a massive region of space that is pulling everything around it towards itself, including our Milky Way galaxy. While we still have much to learn about the Great Attractor, we can be sure that studying this mysterious force will continue to provide new insights into the workings of our universe.

A Celestial Light Show
The Perseid Meteor Shower

Have you ever looked up at the sky at night and seen a shooting star? Shooting stars are not stars but meteors, pieces of rock that enter Earth's atmosphere and burn up, creating a bright streak in the sky. But what if I told you you could see dozens of shooting stars in just one hour?

That happens during the Perseid meteor shower, one of the most spectacular celestial events visible from Earth. The Perseids happen every year around mid-August, and they are caused by the Earth passing through a trail of debris left behind by the comet Swift-Tuttle.

The Perseid meteor shower is named after the constellation Perseus, as the meteors appear to come from the direction of this constellation. But you don't need to know anything about constellations to enjoy the show – just find a dark place away from city lights, lie on a blanket or lawn chair, and look up at the sky.

During the peak of the Perseids, which usually happens around August 12-13, you can see up to 60 or 70 meteors per hour, which means you can see a new shooting star every minute. It's like a fireworks show but in the sky!

What's amazing about the Perseids is the speed at which the meteors travel. They enter the atmosphere at about 60 kilometers per second or 216,000 kilometers per hour! That's more than 50 times faster than a bullet from a gun.

The Perseids have been observed for more than 2,000 years, and they have inspired many legends and stories. In ancient times, people believed that shooting stars were omens of good or bad luck or messages from the gods. Nowadays, we know that shooting stars are just a natural phenomenon, but they still fill us with wonder and awe.

So if you can see the Perseids, don't miss it! It's a free, magical, and unforgettable experience that connects us with the vastness and beauty of the universe.

A Mysterious Explosion in Siberia
The Tunguska Event

In the early morning of June 30, 1908, something strange happened in a remote area of Siberia, now Russia. A massive explosion rocked the region, with force estimated to be equivalent to 10-15 megatons of TNT, over a thousand times more powerful than the atomic bomb dropped on Hiroshima. The shockwave from the blast was so strong that it knocked people off their feet and shattered windows hundreds of miles away.

The cause of this massive explosion remained a mystery for many years, but scientists have now pieced together what happened. It is believed that a small asteroid, estimated to be about 60 meters wide, entered the Earth's atmosphere and disintegrated, creating a shockwave that devastated an area of over 2000 square kilometers.

The location of the explosion was so remote that it took many years for scientists to investigate the site and piece together what happened. When scientists finally arrived in the area, they found a landscape of destroyed trees flattened for miles. The pattern of destruction was centered around a large area completely devoid of trees, indicating that the asteroid likely disintegrated in the air, releasing a tremendous amount of energy.

Scientists have been studying the Tunguska event for over a century, using satellite imagery, computer models, and other tools to understand better what happened. They now believe that the asteroid was likely made of rock and traveled about 33,500 miles per hour.

When it entered the Earth's atmosphere, the air pressure caused it to break apart, releasing an explosion that devastated the region.

While the Tunguska event was a devastating natural disaster, it has also provided scientists with important insights into the potential risks of asteroid impacts. In recent years, there have been increased efforts to track and monitor asteroids to identify potential threats and develop ways to prevent them from striking the Earth.

As we continue to explore the cosmos, events like the Tunguska explosion remind us of the universe's power and unpredictability.

A Faint Glow in the Night Sky
The Zodiacal Light

Have you ever looked up at the night sky and noticed a faint glow stretching upward from the horizon? The Zodiacal Light is one of the most fascinating sights that astronomers can observe.

The Zodiacal Light is a triangular glow visible in the night sky after sunset or before sunrise. It appears to rise from the horizon and extend upward like a cone. This glow is caused by the reflection of sunlight off of countless tiny dust particles scattered throughout our solar system.

These dust particles were left over from the formation of our solar system more than 4 billion years ago. They are found in a disk-shaped region that extends from the sun to beyond the orbit of Jupiter, known as the zodiacal cloud or zodiacal dust cloud. These particles are so small that sunlight can affect them, and the solar wind constantly blows around them.

When we see the Zodiacal Light, we look at the sunlight reflecting off these tiny dust particles. The reflection causes a faint glow that can be seen in the sky. The glow is most visible in areas with little to no light pollution and can be seen from the northern and southern hemispheres.

The Zodiacal Light is named after the zodiac, the twelve constellations that lie along the ecliptic, the path the sun takes through the sky over a year. The Zodiacal Light is most visible in March and April in the northern hemisphere and September and

October in the southern hemisphere. During these times, the ecliptic is at a steep angle to the horizon, making it easier to see the glow in the sky.

Observing the Zodiacal Light can be a fun and rewarding experience for amateur astronomers. To see it:

- Find a location away from city lights with a clear horizon view.

- Look for the glow after sunset or before sunrise when the sky is dark enough to see.

- Allow your eyes to adjust to the darkness for at least 20 minutes to see the faint glow more clearly.

The Zodiacal Light is best seen during the new moon when the sky is darkest.

In conclusion, the Zodiacal Light is a fascinating phenomenon in our solar system. It's caused by the reflection of sunlight off of countless tiny dust particles scattered throughout the zodiacal cloud. It's a beautiful sight that can be observed with the naked eye from a dark location, and it's a reminder of the vastness and complexity of our universe.

A Dark Nebula in Orion
The Horsehead Nebula

Have you ever gazed up at the night sky and spotted a constellation called Orion? Orion is one of the most recognizable constellations in the sky and is home to many fascinating objects, including a nebula known as the Horsehead Nebula.

The Horsehead Nebula is a dark nebula located in the constellation Orion. It is named after its distinctive shape, which resembles a horse's head. The Horsehead Nebula is one of the most well-known nebulae in the sky and a favorite among amateur astronomers.

A nebula is a cloud of gas and dust in space. These clouds are the birthplaces of stars and can be seen throughout the galaxy. The Horsehead Nebula is part of a larger complex of gas and dust known as the Orion Molecular Cloud Complex. This complex is a hotbed of star formation, and it contains many other famous objects, such as the Orion Nebula and the Flame Nebula.

The Horsehead Nebula is a particularly interesting object because it is dark. Unlike other nebulae, which are visible due to the light emitted by the gases they contain, the Horsehead Nebula is visible because it blocks the light from the stars behind it. The nebula is made up of cold, dense gas and dust, which absorbs the light from the stars and prevents it from reaching us here on Earth.

The Horsehead Nebula is located about 1,500 light-years away from Earth and is relatively small, measuring only about three light-years across. Despite its small size, it is an incredibly complex and detailed object, best observed through a telescope.

The Horsehead Nebula was first discovered in 1888 by Williamina Fleming, a Scottish astronomer working at the Harvard College Observatory. Since then, it has become a favorite among amateur astronomers and astrophotographers, who often capture stunning nebula images using long-exposure photography.

Studying the Horsehead Nebula and other nebulae like it is an important part of understanding how stars form and evolve. By studying the gas and dust in these clouds, astronomers can learn more about the conditions that give rise to new stars and gain insights into the life cycles of stars.

In conclusion, the Horsehead Nebula is a fascinating object in the constellation Orion. It is a dark nebula, which is visible because it blocks the light from the stars behind it. Despite its small size, it is an incredibly complex and intricate object that has captured the imaginations of astronomers and stargazers for over a century. By studying the Horsehead Nebula and other nebulae like it, astronomers can better understand how stars form and evolve.

A Spectacular Spiral Galaxy
The Sombrero Galaxy

The Sombrero Galaxy, also known as Messier 104, is a spectacular spiral galaxy located approximately 28 million light-years from Earth in the constellation Virgo. It is called the Sombrero Galaxy because it has a dark, circular band of dust and gas that gives it the appearance of a wide-brimmed hat or a Mexican sombrero.

The Sombrero Galaxy has a diameter of about 50,000 light-years, roughly half the size of the Milky Way. It has a bright nucleus surrounded by a bulge of older stars and a large, flattened disk of gas and dust that contains younger stars. The disk is surrounded by a halo of globular clusters, tightly packed groups of ancient stars that orbit the galaxy.

One of the most interesting features of the Sombrero Galaxy is its supermassive black hole, which has a mass equivalent to about 1 billion suns. This black hole is surrounded by a swirling disk of gas and dust pulled in by the black hole's gravity. As the gas and dust fall into the black hole, they heat up and emit bright X-rays and other forms of radiation that telescopes on Earth can detect.

The Sombrero Galaxy was first discovered by the French astronomer Pierre Méchain in 1781. It was later added to the famous Messier Catalog of celestial objects by the French astronomer Charles Messier. Today, the Sombrero Galaxy is a popular target for amateur astronomers and professional astronomers alike, who study it to learn more about the formation and evolution of galaxies.

In summary, the Sombrero Galaxy is a spectacular spiral galaxy located in the constellation Virgo, approximately 28 million light-years away from Earth. It has a bright nucleus, a bulge of older stars, a large disk of gas and dust containing younger stars, and a halo of globular clusters. Its supermassive black hole, which has a mass equivalent to about 1 billion suns, is surrounded by a swirling disk of gas and dust that emits bright X-rays and other forms of radiation. The Sombrero Galaxy is an important object of study for astronomers who want to learn more about the formation and evolution of galaxies.

The First US-Soviet Space Mission
The Apollo-Soyuz Test Project

Now we will talk about a historical event in space – the Apollo-Soyuz Test Project, which was the first time the United States and the Soviet Union worked together on a space mission.

In the 1960s and 1970s, the United States and the Soviet Union were in a Cold War, a time of political tension between the two superpowers. One of the ways that they competed was through space exploration. The Soviet Union launched the first satellite, sent the first human into space, and had the first spacewalk. The United States, on the other hand, landed humans on the Moon. But despite their differences, they both shared a passion for space exploration and the desire to push the boundaries of what was possible.

The Apollo-Soyuz Test Project was a joint mission to test how well the two countries could work together in space. The mission comprised two parts: the American part was called Apollo, and the Soviet part was called Soyuz. On July 15, 1975, the Apollo spacecraft was launched from Florida in the United States, while the Soyuz spacecraft was launched from Baikonur in the Soviet Union. The two spacecraft met and docked in orbit on July 17.

The two crews worked together to experiment, exchange gifts, and learn from each other. They even ate meals together! That was an important moment for international cooperation in space. It showed that even two countries with very different political systems and ideologies could work together toward a common goal.

The Apollo-Soyuz Test Project also had important scientific and technical achievements. It was the first time two spacecraft from different countries had docked in space. That was an important step towards future international cooperation in space, such as the International Space Station. The mission also tested new docking equipment and communication systems, which helped to improve future space missions.

After a few days of joint activities, the spacecraft undocked and returned to Earth. The Apollo spacecraft landed in the Pacific Ocean, while the Soyuz landed in Kazakhstan. The crews were greeted as heroes in their respective countries, and the mission was seen as a symbol of hope for international cooperation.

In conclusion, the Apollo-Soyuz Test Project was a significant moment in space exploration history, demonstrating that even during political tension, countries could come together for a common purpose. The mission paved the way for future international cooperation in space and contributed to important scientific and technical advancements.

The Pillars of Creation
The Eagle Nebula

The Eagle Nebula, also known as Messier 16, is a stunning region of space located approximately 7,000 light-years away from Earth. This nebula is famous for its "Pillars of Creation," which are towering columns of gas and dust that are forming new stars.

The Eagle Nebula is a young star-forming region with the brightest and most massive stars only a few million years old. These stars are incredibly hot and emit intense ultraviolet radiation, which ionizes the surrounding gas, creating the characteristic glow of the nebula.

The Pillars of Creation are the most striking feature of the Eagle Nebula. They are massive columns of gas and dust, each several light-years long. These pillars create new stars as the gas and dust within them collapse under their own gravity. The intense radiation from nearby stars has carved out the distinctive shape of the pillars, eroding the denser parts of the gas and dust while leaving behind the denser, star-forming regions.

In 1995, the Hubble Space Telescope captured an iconic image of the Pillars of Creation. This image showed the pillars in stunning detail and has become one of the most famous astronomical images ever. However, the Pillars of Creation are not eternal. The intense radiation from nearby stars is eroding them, and some scientists believe they may have already been destroyed, with the light from their destruction yet to reach us.

The Eagle Nebula is also home to several other fascinating features, including young star clusters and dark nebulae, regions of space where dense dust clouds obscure the light from behind them. Observing these features can give scientists insights into the formation and evolution of stars and galaxies.

Overall, the Eagle Nebula is a beautiful and fascinating region of space that provides a glimpse into the complex processes that shape our universe. Its Pillars of Creation have become an iconic symbol of the majesty and wonder of the cosmos.

The Center of Our Solar System
The Sun

Now, we will talk about the center of our solar system - the Sun!

The Sun is a giant, glowing ball of gas and is the closest star to Earth. It is located at the center of our solar system and is responsible for providing heat and light to all the planets orbit around it.

The Sun is so massive that it makes up about 99.86% of the solar system's total mass. It is around 109 times larger than the Earth and has a diameter of about 1.39 million kilometers!

The Sun is also extremely hot, reaching up to 15 million degrees Celsius at its core. Despite its incredibly high temperatures, the Sun mostly comprises hydrogen (about 73%) and helium (about 25%). These two elements undergo nuclear fusion, a process in which the atoms combine and release tremendous energy in the form of heat and light.

The energy released by the Sun travels through space in the form of light and takes about 8 minutes to reach Earth. This energy is essential for life on Earth, as it helps plants grow, allows us to see, and provides warmth.

The Sun also creates space weather, which can affect our planet. When the Sun's magnetic field lines become twisted and distorted, they can cause massive eruptions of plasma, known as coronal mass ejections (CMEs), and release solar flares. These events can create beautiful auroras but can also interfere with communication systems, satellites, and power grids.

Scientists study the Sun using a variety of instruments and techniques, such as telescopes, satellites, and ground-based observatories. By studying the Sun, we can learn more about our planet and other objects in the solar system.

In conclusion, the Sun is the most important object in our solar system, providing us with heat, light, and energy. It is a fascinating and essential subject for scientists, and there is still much to learn about this giant ball of gas.

The Pleiades Star Cluster
The Seven Sisters

Let's learn about a beautiful star cluster in the night sky called the Pleiades, also known as the Seven Sisters.

The Pleiades is a group of stars located in the constellation Taurus. To the naked eye, it appears as a small cluster of stars that twinkle together in the sky. However, when viewed through a telescope, the cluster reveals a stunning array of bright, hot, and blue stars.

According to Greek mythology, the Seven Sisters were the daughters of Atlas and Pleione, and they were transformed into stars to escape from the advances of Orion. The Pleiades have been known and admired since ancient times, and many cultures worldwide have mentioned them.

The Pleiades star cluster is relatively young, estimated to be only around 100 million years old, considered very young compared to our Sun's 4.6 billion-year age. The cluster is also relatively close to us, at a distance of about 440 light-years away. The cluster contains over 1,000 stars, with the brightest and most massive stars shining brightly and dominating the cluster's appearance.

The Seven Sisters is a beautiful sight to see and has scientific significance. Astronomers use the Pleiades as a benchmark to calibrate the distance to other objects in space, and they also study the cluster to understand the life cycle of stars better.

In conclusion, the Pleiades, or the Seven Sisters, is a beautiful star cluster in the night sky that people have admired for centuries. It is relatively young and close to us and has scientific importance for astronomers. Next time you look up at the sky, try to find the Seven Sisters and appreciate their beauty and significance in our universe.

Visitors from the Outer Solar System
The Comets

Comets are one of the most fascinating objects in our solar system. These icy bodies come from the outer edges of the solar system, and their journey through space can be beautiful and unpredictable.

Comets comprise three parts: the nucleus, the coma, and the tail. The nucleus is the solid center of the comet, made up of ice, dust, and rock. When a comet gets close to the sun, the ice on the nucleus starts to vaporize, creating a coma, a cloud of gas and dust surrounding the nucleus. Finally, as the comet moves closer to the sun, the solar wind pushes the gas and dust away from the nucleus, creating a tail that can stretch for millions of kilometers.

The study of comets is important for scientists because they can tell much about the early solar system. Comets are thought to be leftover building blocks from the solar system's formation, so studying them can help us understand how our planet and others formed.

Comets have been observed and recorded for thousands of years, and they often have been seen as harbingers of important events. In the 11th century, Halley's Comet was seen as a bad omen before the Battle of Hastings in 1066. And in the 14th century, the appearance of a bright comet was seen as a sign of the Black Death that swept through Europe.

Today, we have sent spacecraft to study comets up close. In 2014, the European Space Agency's Rosetta spacecraft rendezvoused with the comet 67P/Churyumov-Gerasimenko and even landed a small probe, Philae, on its surface. This mission gave us unprecedented insight into the composition and behavior of comets.

In conclusion, comets are fascinating objects in our solar system that have captured the imagination of people for thousands of years. Their study has helped us understand the early solar system and how our planet and others formed. And with the help of modern spacecraft, we are learning more about these icy visitors from the outer solar system than ever before.

High-Energy Particles from Space
The Cosmic Rays

Now, we will talk about a fascinating phenomenon called cosmic rays. These are particles that come from space and travel at incredibly high speeds. Let's explore what they are, where they come from, and why they matter.

So, what exactly are cosmic rays? They comprise high-energy particles, such as protons, electrons, and atomic nuclei. They come from outer space and travel at almost the speed of light. They can pass through solid objects, including our bodies, without harming us. Cosmic rays constantly bombard us as they pass through the Earth's atmosphere.

Cosmic rays come from various sources, including the sun, other stars, and even distant galaxies. They are produced by supernovae, which are massive explosions that occur when a star runs out of fuel and collapses. These explosions can create cosmic rays with energies millions of times greater than anything we can produce on Earth.

Scientists have been studying cosmic rays for many years and have discovered some fascinating things about them. For example, cosmic rays can tell us about the universe's composition and galaxy's structure. They can also help us understand the behavior of particles at very high energies.

One of the biggest mysteries surrounding cosmic rays is their origin. Scientists still don't know where all cosmic rays come from or how they are produced. It is an area of active research, and discoveries are always being made.

Despite the many unknowns, cosmic rays are incredibly important for understanding the universe. They allow us to study particles and energies that are impossible to create on Earth. They also play a role in the universe's formation, and they could even affect life on Earth.

In conclusion, cosmic rays are high-energy particles that come from space and travel at almost the speed of light. Supernovae and other astrophysical processes produce them, and they can tell us a lot about the universe. While we still don't know much about cosmic rays, scientists continue to make discoveries and expand our understanding of these magnetic particles.

Our Satellite Galaxies
The Magellanic Clouds

The Magellanic Clouds are two small irregular dwarf galaxies located near our Milky Way galaxy. They are named after Ferdinand Magellan, the Portuguese explorer who first saw them during his voyage around the world. The Large Magellanic Cloud (LMC) is about 160,000 light-years away, while the Small Magellanic Cloud (SMC) is about 200,000 light-years away.

Even though the Magellanic Clouds are small compared to our Milky Way galaxy, they are still among the brightest and most recognizable features in the southern hemisphere's night sky. They contain billions of stars, gas, and dust and are home to several star-forming regions.

The Magellanic Clouds are also important to astronomers because they are some of the closest galaxies to us, making them excellent objects to study. By observing them, we can learn more about the evolution of galaxies and how they form and change over time.

One of the most interesting things about the Magellanic Clouds is their interaction with our Milky Way galaxy. The Magellanic Clouds are moving towards the Milky Way and will eventually merge with our galaxy in about two billion years. This event will cause a burst of star formation, dramatically changing the Milky Way's structure.

In addition, the Magellanic Clouds are also responsible for creating some of the most stunning objects in the night sky. One example is the Tarantula Nebula, which is located in the LMC. This nebula is a giant star-forming region, one of the largest and brightest nebulae known. It contains many hot, young stars and is a popular target for amateur and professional astronomers.

In conclusion, the Magellanic Clouds are fascinating satellite galaxies that have captured the attention of astronomers and stargazers alike. They are home to billions of stars and some of the most spectacular objects in the night sky. As we continue to study them, we will better understand the universe and our place in it.

A Mysterious Storm
The Vortex of Saturn

Let's talk about a strange phenomenon on one of the most spectacular planets in our solar system: Saturn. Specifically, we will be looking at the mysterious vortex found at the planet's north pole.

Saturn is the sixth planet from the sun and is known for its stunning rings, which are dust, ice, and rocks. But that's not all that's amazing about this gas giant. Saturn also has a unique weather system that includes massive storms, such as the Great White Spot, which can be seen from Earth.

One of the most fascinating features of Saturn's weather system is the vortex located at its north pole. A vortex is a swirling mass of air or liquid, similar to a whirlpool. In the case of Saturn's vortex, the swirling is caused by the planet's powerful winds, which can reach speeds of up to 1,100 miles per hour (1,800 kilometers per hour)!

But what makes this vortex so mysterious is its hexagonal shape. That's right – the swirling mass of gas is shaped like a six-sided polygon. Scientists have been trying to figure out what causes this unusual shape for years, but they haven't come up with a definitive answer.

One theory is that the hexagonal shape is caused by the planet's rotation and the way its atmosphere moves. Another theory suggests that the shape is created by a series of standing waves that form in the planet's atmosphere. It's also possible that the shape is caused by the planet's magnetic field, which interacts uniquely with the atmosphere.

Whatever the cause of the hexagonal vortex, it's an amazing sight to behold. Scientists have been studying the phenomenon for years, using data from NASA's Cassini spacecraft, which orbited Saturn from 2004 to 2017. They've also used computer simulations to understand the vortex dynamics better.

In recent years, scientists have discovered similar hexagonal shapes on other planets, such as Jupiter and Neptune. But the vortex on Saturn remains the most famous and mysterious.

So, there you have it, curious kids! The vortex at Saturn's north pole is a fascinating and mysterious phenomenon caused by the planet's powerful winds and shaped like a hexagon. While scientists are still trying to understand exactly what causes this unique feature, we can all appreciate its beauty and wonder.

The Next Frontier
The James Webb Space Telescope

The James Webb Space Telescope is an incredible human engineering and ingenuity feat. It is set to be the successor to the Hubble Space Telescope and is designed to observe the universe in ways that were never possible.

The James Webb Space Telescope is named after James E. Webb, who served as the second administrator of NASA from 1961 to 1968. It is a joint project of NASA, the European Space Agency (ESA), and the Canadian Space Agency (CSA). The telescope is set to launch in 2021, and it will be stationed at a special point in space called the second Lagrange point, which is located about 1.5 million kilometers (nearly 1 million miles) from Earth.

The telescope is enormous, standing at about 6.5 meters (over 21 feet) tall and weighing around 6.2 metric tons (over 13,000 pounds). It is designed to detect light from some of the first galaxies that formed in the universe, and it is also capable of observing planets around other stars. The telescope has several instruments, including a near-infrared camera, a mid-infrared instrument, and a near-infrared spectrograph.

One of the most exciting features of the James Webb Space Telescope is its ability to study the atmospheres of exoplanets, which are planets outside of our solar system. By analyzing the light that passes through a planet's atmosphere, scientists can determine what gases are present and possibly even detect signs of life.

However, the James Webb Space Telescope has not been without its challenges. Its development has taken many years and has experienced significant delays and cost overruns. But despite these setbacks, scientists and engineers are determined to see it through to launch and are eager to see what discoveries it will bring.

In conclusion, the James Webb Space Telescope is an incredible scientific instrument that has the potential to revolutionize our understanding of the universe. It represents the next frontier in space exploration and promises to unlock new insights into the origins and nature of the cosmos.

A Cosmic Structure
The Great Wall of Galaxies

The universe is full of wonders that continue to amaze us, and one of the most spectacular sights in the cosmos is the Great Wall of Galaxies. It is a vast cosmic structure, spanning more than half a billion light-years across, and consists of thousands of galaxies held together by gravity.

The Great Wall of Galaxies is located in the direction of the constellation Hercules, one of the largest structures in the universe that we know of. It was first discovered in the late 1980s by astronomers studying galaxies' distribution in the universe.

At the time, the prevailing theory was that the universe was homogeneous, meaning it looked the same in all directions. However, the discovery of the Great Wall of Galaxies shattered this idea and showed that the universe is clumpy and uneven.

The Great Wall of Galaxies is a vast network of galaxies that stretches across the cosmos. It is made up of several smaller structures that are held together by gravity, and it is so large that it takes light more than a billion years to travel from one end to the other.

Scientists believe that the Great Wall of Galaxies formed billions of years ago, during the early stages of the universe when it was still very young. At this time, the universe was filled with gas and dust, and galaxies were forming rapidly.

As these galaxies formed, they began to clump together, and over time, these clumps grew larger and larger, eventually forming the massive structure that we see today.

The Great Wall of Galaxies is an incredible sight to behold, and it gives us a glimpse into the vastness and complexity of the universe. While we still have much to learn about this amazing structure, it reminds us how much there is still to discover and explore in the cosmos.

A Radiation Shield
The Van Allen Belts

Have you ever heard of the Van Allen Belts? They are two regions of intense radiation that surround our planet Earth, and they were named after the scientist who discovered them, Dr. James Van Allen.

The Van Allen Belts were discovered in 1958 by the first American satellite, Explorer 1. The satellite carried a Geiger counter, an instrument that detects radiation. When Explorer 1 was launched into space, the Geiger counter detected something unusual: two regions of intense radiation surrounding our planet.

These two regions of intense radiation are called the Van Allen Belts. They are located in the Earth's magnetosphere, the region of space surrounding our planet where Earth's magnetic field traps charged particles from the sun. The Van Allen Belts are made up of charged particles such as electrons and protons, which are trapped by the Earth's magnetic field and forced to travel in a circular path around the Earth.

The Van Allen Belts are very important for protecting life on Earth. They shield against harmful radiation from space, such as cosmic rays and solar flares. Without the Van Allen Belts, life on Earth would be exposed to dangerous levels of radiation that could cause harm to living things.

However, the Van Allen Belts can also be a problem for satellites and spacecraft. The intense radiation can damage electronic equipment and even cause them to malfunction. Scientists and engineers must design spacecraft and satellites with the Van Allen Belts in mind to ensure they are protected from radiation.

In summary, the Van Allen Belts are two regions of intense radiation that surround our planet Earth, and they play a crucial role in protecting life on Earth from harmful radiation from space. While they can be a problem for spacecraft and satellites, they are a natural shield that makes our planet habitable for living things.

Neutron Stars with a Purpose
The Pulsars

Let us explore the fascinating world of pulsars. These neutron stars emit electromagnetic radiation beams out of their magnetic poles, which are usually not aligned with their rotational axis. It causes the radiation to appear as pulses, hence the name pulsars.

Neutron stars are the collapsed cores of massive stars that have exploded as supernovas. They are incredibly dense, with a mass roughly 1.4 times that of the sun but squeezed into a ball only about 12 miles across. It means they are extremely compact and have strong gravitational and magnetic fields.

Pulsars were first discovered by radio astronomer Jocelyn Bell Burnell in 1967. She was studying radio emissions from quasars when she noticed a periodic and persistent signal. At first, she thought it might be interference from human sources, but she soon realized it was coming from a celestial object. After ruling out other possibilities, she and her advisor, Antony Hewish, concluded that they had discovered a new type of celestial object called pulsars.

Pulsars emit radiation across various wavelengths, from radio waves to X-rays and gamma rays. The radiation is thought to be caused by the acceleration of charged particles in the pulsar's magnetic field, which is incredibly strong due to the star's small size and high rotation rate.

One of the most remarkable properties of pulsars is their incredible stability. The pulses of radiation they emit are incredibly precise and regular, with some pulsars maintaining their frequency for billions of years. It makes pulsars excellent natural clocks and has allowed astronomers to test theories of gravity and even search for gravitational waves.

Pulsars are also important for our understanding of the universe. By studying the properties of pulsars and their radiation, astronomers can learn about the extreme conditions in the cores of massive stars and the structure and properties of the magnetic fields that surround them.

In conclusion, pulsars are incredibly fascinating celestial objects that have helped us expand our understanding of the universe. They have allowed us to test theories of gravity, search for gravitational waves, and learn about the extreme conditions in the cores of massive stars. It's amazing to think that something as small as a neutron star can emit beams of radiation that we can detect from Earth and that studying these pulsars can help us answer some of the biggest questions about the universe.

The Proxima Centauri System
A Close Neighbor with a Potentially Habitable Planet

The Proxima Centauri system is a fascinating area of space located just 4.24 light-years away from our solar system. It is the closest star system to our own and has been the subject of much study and speculation in recent years. The Proxima Centauri system is so interesting because it contains a potentially habitable planet called Proxima B.

Proxima B is an exoplanet that orbits Proxima Centauri, which is a red dwarf star that is smaller and cooler than our own Sun. Proxima b is about 1.3 times the size of Earth, and it orbits its star at a distance that could allow liquid water to exist on its surface. It has led many scientists to speculate that Proxima B could be a potentially habitable world.

Of course, many challenges are associated with studying a planet that is so far away. One of the main challenges is simply getting there. With current technology, traveling to the Proxima Centauri system would take thousands of years, so we have had to rely on other methods to study Proxima B.

One of the main ways we have learned about Proxima B is by observing the star it orbits. When a planet passes in front of its star, as seen from Earth, it causes a slight dimming of the star's light. By carefully measuring this dimming, astronomers can determine the planet's size, orbit, and other properties.

Another way that we have learned about Proxima B is by studying the gravitational interactions between the planet and its star. By carefully observing how the star "wobbles" as the planet orbits it, astronomers can determine the planet's mass and other properties.

While there is still much we don't know about Proxima B and the Proxima Centauri system, it is an exciting area of study that could have important implications for the search for extraterrestrial life. Who knows what other surprises and mysteries await us in this fascinating corner of the universe?

The Future Fate of Our Universe
The Black Dwarf Stars

As you may know, stars are enormous balls of hot gas that shine brightly in the sky. They come in all sizes, from small red dwarfs to giant blue supergiants. The sun, the star at the center of our solar system, is a medium-sized yellow star.

When stars run out of fuel, they start to die. Small stars like red dwarfs can die quietly, becoming dim and cool objects called brown dwarfs. Medium-sized stars like the sun will expand into a red giants before shedding their outer layers to become a small, hot object called a white dwarf. But what happens to larger stars at the end of their lives?

Well, when really big stars run out of fuel, they explode in a massive supernova explosion. The explosion is so bright that it can briefly outshine an entire galaxy! What's left behind after a supernova is either a black hole or a neutron star, depending on how massive the star was, to begin with.

But what happens to white dwarfs, the remnants of smaller stars, after they cool down? That's where black dwarfs come in. A black dwarf is a white dwarf that has cooled down so much that it no longer emits visible light. Instead, it becomes a dark, cold, and dense object.

The thing is, black dwarfs, don't exist yet. The universe is only about 14 billion years old, and it takes much longer than that for a white dwarf to cool down enough to become a black dwarf. Scientists estimate that it could take trillions of years for the first black dwarfs to form!

So, what does all of this mean for the future of our universe? As more and more stars run out of fuel and become white dwarfs, they will eventually cool down to become black dwarfs. As the number of black dwarfs increases, the universe will become darker and colder.

Eventually, trillions of years from now, the last stars will die, leaving behind only black dwarfs and black holes. The universe will become dark and lonely, with no new stars forming and no light to guide the way. It is known as the "heat death" of the universe, a fate that astronomers believe is inevitable.

So, there you have it, curious kids! Black dwarfs may not exist yet, but they are an important part of the future of our universe. Who knows what other mysteries the universe has in store for us as we continue to explore and learn more about the cosmos?

The Cosmic Clouds of Gas and Dust
The Nebulae

Let us talk about one of the universe's most spectacular and mysterious objects: Nebulae! Nebulae are vast clouds of gas and dust scattered throughout the cosmos. They are known for their breathtaking colors and stunning shapes, making them some of the most beautiful objects in the night sky.

There are many types of nebulae, each with its unique characteristics. One of the most famous types is the H II region nebula, a massive cloud of ionized hydrogen gas often associated with young stars. These nebulae can be very colorful, ranging from bright red to blue, and they are often located in the spiral arms of galaxies like our Milky Way.

Another common type of nebula is the planetary nebula, a cloud of gas that forms around a dying star. Despite their name, these nebulae have nothing to do with planets! Instead, they are called planetary nebulae because early astronomers thought they looked like planets through their telescopes. Planetary nebulae can have a wide range of shapes, from simple spheres to intricate patterns, and their colors can also vary widely.

Finally, we have the dark nebulae. As the name suggests, these clouds are dark and almost invisible, as they block the light from the stars behind them. However, they are not completely devoid of light. They often appear as faint silhouettes against the background of brighter stars or nebulae.

Nebulae are not just pretty to look at, though. They play a critical role in the life cycle of stars. When a massive star reaches the end of its life, it explodes in a supernova, scattering its elements throughout space. These elements are then incorporated into the nebulae, eventually becoming the building blocks of new stars and planets.

The study of nebulae has been essential to our understanding of the universe. Through observations of nebulae, astronomers have learned about the life cycles of stars, the composition of the universe, and even the structure of our own galaxy. Nebulae are also a source of fascination and inspiration for people worldwide, serving as a reminder of the beauty and complexity of the cosmos.

In conclusion, nebulae are some of the universe's most fascinating and beautiful objects. From their stunning colors and intricate shapes to their critical role in the formation of stars and planets, they are a reminder of the vastness and complexity of the cosmos. So, the next time you gaze up at the night sky, take a moment to appreciate the cosmic clouds of gas and dust that make it such a spectacular sight.

The CMB Cold Spot

A Strange Anomaly in the Cosmic Microwave Background

The Cosmic Microwave Background (CMB) is the oldest light in the universe, dating back to just 380,000 years after the Big Bang. Scientists study this radiation to learn about the early universe and its evolution. The CMB appears as a faint, uniform glow in the sky, but it also has small temperature variations that reveal information about the universe's structure.

However, one peculiar feature in the CMB is known as the CMB Cold Spot. This region appears colder than the rest of the CMB by about 0.00015 degrees Celsius, which may not seem like a lot, but it is a significant deviation in the context of the CMB.

The Cold Spot was first discovered in 2004 by a team of astronomers using data from the Wilkinson Microwave Anisotropy Probe (WMAP), a satellite designed to measure the CMB. The team found that the Cold Spot was much larger than they expected, covering an area of about 70 degrees in the sky, or about 30 times the size of the full moon.

Since its discovery, scientists have been trying to understand the nature of the Cold Spot. One theory is that it may be caused by a supervoid, a space region with fewer galaxies and less matter than average. To account for the Cold Spot's temperature deviation, the supervoid would have to be extremely large, about 1.8 billion light-years across.

Another theory is that the Cold Spot is a remnant of a collision between our universe and another universe, known as the multiverse theory. According to this theory, the Cold Spot may result from a gravitational tug-of-war between our universe and another universe, which would have left an imprint in the CMB.

However, many unanswered questions remain about the Cold Spot, and more research is needed to determine its true nature. But this mysterious feature in the CMB intrigues astronomers and may lead to new discoveries about the universe and its evolution.

A Cosmic Void in Space
The Eridanus Supervoid

Have you ever heard of a place where there seems to be nothing, not even stars or galaxies? It might sound like science fiction, but it is a real place in space called the Eridanus Supervoid.

The Eridanus Supervoid is an enormous empty region of space located in the constellation Eridanus, which is about 1.3 billion light-years away from us. It is known as a supervoid because it is much larger than a typical void in space.

Scientists first discovered the Eridanus Supervoid in the late 2000s while studying a map of the cosmic microwave background radiation, the oldest light in the universe. This radiation is leftover from the Big Bang and can tell us a lot about the universe's structure.

When scientists looked at this map, they noticed a very large and unusually cold spot in the cosmic microwave background radiation that corresponded with the location of the Eridanus Supervoid. This cold spot is thought to be caused by the lack of matter in the region, which means there is less gas and dust to create new stars and galaxies.

The Eridanus Supervoid is about one billion light-years across, which means it is almost one-tenth of the observable universe's size. To give you an idea of how big that is, if you traveled across the void at the speed of light, it would take you one billion years to cross it!

Scientists are still studying the Eridanus Supervoid to try and understand why it is so empty. One theory is that the void was created by the gravitational pull of a nearby supercluster of galaxies, which caused matter to be pulled away from the region. Another theory is that it is simply a statistical fluke and that the lack of matter in the region is just a coincidence.

The Eridanus Supervoid is a fascinating and mysterious place in space that reminds us just how little we know about the universe. Who knows what other strange and incredible discoveries await us as we continue to explore the cosmos?

The Cosmic Web
The Large-Scale Structure of the Universe

The universe is vast and full of mysteries, but scientists have studied it for centuries, trying to understand its structure and how everything fits together. One of the most fascinating discoveries of modern astronomy is the cosmic web, the universe's large-scale structure.

The cosmic web is a complex network of clusters, filaments, and voids that spans the entire universe. It comprises billions of galaxies, each containing billions of stars. It is held together by dark matter, a mysterious substance that we cannot see or detect directly but whose presence we can infer from its gravitational effects.

The cosmic web is believed to have formed shortly after the Big Bang, as matter began to clump together and gravity began to pull it into large structures. Over time, these structures grew and evolved, forming the intricate web of galaxies and filaments we see today.

The cosmic web is not evenly distributed throughout the universe. Instead, it is characterized by large voids, regions of space with few or no galaxies, filaments, long chains of galaxies, and gas stretching across space. These filaments intersect at points called nodes, where clusters of galaxies form.

One of the most important tools for studying the cosmic web is the cosmic microwave background radiation (CMB). The faint glow of radiation left over from the Big Bang permeates the entire universe. By studying the patterns in the CMB, scientists can learn about the distribution of matter in the universe and how it has evolved.

The cosmic web is fascinating from a scientific perspective and has important implications for our understanding of the universe and our place within it. It shows us that the universe is not just a random collection of stars and galaxies but has a structure and organization that we are only beginning to understand.

In conclusion, the cosmic web is a vast and complex structure that spans the entire universe, comprising clusters, filaments, and voids. It's held together by dark matter and is characterized by large voids and intersecting filaments forming nodes where galaxy clusters are found. Studying the cosmic web is important for understanding the universe and our place within it.

AUTHOR'S NOTE

Dear Reader,

Thank you for taking the time to explore the wonders of the universe through my book, Marvels of Space: 50 Amazing Stories of the Universe. I hope it has expanded your knowledge and curiosity about the cosmos.

As a science enthusiast and researcher, I have always been fascinated by the mysteries of space, and it brings me great joy to share my passion with others through this book. I believe that the more we learn about the universe, the more we appreciate its beauty and complexity.

If you enjoyed reading this book, I encourage you to explore more of my work and other astronomy and space exploration books. The universe is vast, and there is always more to discover.

Thank you again for joining me on this journey through space.

Sincerely,

Sam Blackwood

www.ingramcontent.com/pod-product-compliance
Lightning Source LLC
Chambersburg PA
CBHW070358220526
45467CB00001B/432